D0590786

IRELAND'S BIRD LIFE

A World of Beauty

IRELAND'S BIRD LIFE

A World of Beauty

EDITORS
Matt Murphy
Susan Murphy

PHOTOGRAPHER
Richard Mills

TEXT
Richard Lansdown

A Sherkin Island Marine Station Publication

Foreword

It gives me great pleasure to introduce this most attractive photographic publication on Irish birds. It should be a highly prized possession of all those interested in Ireland's rich natural heritage as it portrays to us the rich diversity of Irish bird life, a great part of our natural inheritance.

For anybody familiar with its sister volume "Ireland's Marine Life - A World of Beauty" this striking and inspirational compendium of Ireland's bird life will come as no surprise. It is just another example of the high quality work and dedicated professionalism we have come to know and expect from Sherkin Island Marine Station. Through its newspaper "Sherkin Comment" and these books the Marine Station makes major contributions to our understanding and appreciation of our flora and fauna.

I wish to pay tribute to the editors Matt and Susan Murphy and extend my congratulations to all those involved in this excellent work. The brilliant photography is by Richard Mills, recognised as one of the finest wildlife photographers in Europe.

I hope that you derive great pleasure from this book and share in my appreciation and concern for the conservation of Ireland's bird life.

Noel Dempsey T.D.,
Government Chief Whip and
Minister for State at the Department of Finance.

This book is dedicated to
Matty, Michael, Susan, Mark,
Robbie, Peter and Audrey,
who gave so much of their growing
years to building Sherkin Island
Marine Station.

Published by SHERKIN ISLAND MARINE STATION PUBLICATIONS
First published May 1994

British Library Cataloguing-in-Publication Data.
A catalogue record for this book is available from the British Library.

ISBN: 1 870492 80 3

Design & typesetting: Susan Murphy, Sherkin Island Marine Station, Co. Cork, Ireland.
Colour Separations by: Academy Litho Ltd., Dublin, Ireland.
Printed by: City Printing Works Ltd., Victoria Cross, Cork, Ireland.
Binding by: Library Binding, Dublin, Ireland.

ACKNOWLEDGEMENTS

Sherkin Island Marine Station would like to acknowledge the assistance of a number of people who made a special contribution to this book.

Thanks to John O'Halloran; Michael Herley; Oscar Merne; Richard Mundy; Noel and Michael Dunlea, Paper Sales Ltd., Cork; Denis McSweeney PhotoShop, Cork; the Courtney Family, City Printing Works, Cork; Pat Walsh; Colm Breathnach, Secretary, An Buanchoiste Téarmaíochta (The Permanent Terminology Committee of the Department of Education).

The Irish names in this book are taken from: *"Ainmneacha Plandaí agus Ainmhithe - Flora and Fauna Nomenclature"*

Introduction

For many years we at Sherkin Island Marine Station have admired the photography of Richard Mills whose work has graced many publications in Ireland and abroad. These include the *Cork Examiner* newspaper, where he is a staff photographer. Richard has visited our Station over the years and each time we have been fascinated and in awe at his patience and attention to detail.

The photographs we have chosen from his vast collection will, we are certain, make many people wish that they too could take such wonderful photographs. They will, like us, be envious of his great talents but will realise that every craft has a chosen few who surpass the norm. Richard is one of them.

This book is not for the expert birdwatcher or photographer and is by no means a complete guide to Irish birds. It is for the many hundreds of thousands who, like us, know little or nothing about these wonderful creatures. It is hoped that the book will encourage many to take up birdwatching as a hobby.

What is even more important is that people will realise that wildlife habitats must be protected as urban Ireland expands into the countryside. In particular we hope planners of new roadways will replace at least some of the many miles of hedgerows they destroy each year. Too many of Ireland's wildlife habitats have been lost and, unless the slide is stopped, our children and grandchildren will not enjoy the wonderful bird life which frequents our skies, some of which we feature in this book.

We would like to thank sincerely the following sponsors for the generous support we received for the book:

Bord Fáilte

Cork Examiner Publications Ltd.

Dept. of the Environment

Henkel Ireland Ltd.

National Parks & Wildlife Service of the Office of Public Works

Janssen Pharmaceuticals Ltd.

Schering-Plough (Brinny)

SmithKline Beecham

Údarás na Gaeltachta

Yamanouchi Ireland

Little grebe or Dabchick

Tachybaptus ruficollis

Spágaire tonn

The little grebe is a familiar sight on reed fringed pools and rivers where it builds a floating nest of reeds and pondweeds. When it leaves the nest, the adult covers its white eggs with leaves from the nest to avoid detection by predators.

⇩

Fulmar

Fulmarus glacialis

Fulmaire

⇧

With their stiff-winged gliding flight, their close relationship to shearwaters and albatrosses is obvious. Here a fulmar rides upwelling wind currents above the colony.

Fulmars breed on grassy ledges on inaccessible cliffs. Although some are present around the colonies throughout much of the year, they spend a lot of their adult lives at sea. This photograph shows a group sitting at the top of the cliff above their colony, squabbling.

⇩

Storm petrel

Hydrobates pelagicus

Guairdeall

⇧

Storm petrels resemble house martins in their shape and white rump. However, they are a truly oceanic species, rarely seen inland except in very strong winds. They will sometimes gather in groups of thousands to feed over up-welling currents along the edge of the continental shelf. Storm petrels are very weak on land and unable to defend themselves against predators. Therefore, they wait offshore until dark before moving in to replace their partner incubating their single egg.

Gannet

Sula bassana

Gainéad

⇧

With their striking yellow head, white body and black "dipped in ink" wingtips, gannets present a dramatic sight and can easily be recognised at great distance.

Gannets have a similar soaring flight to that of fulmars, to save energy when covering large distances at sea. Here an adult bird hangs in the air above the colony on the Saltee Islands.

⇦

Gannets nest in huge ancestral colonies on rocky sea cliffs and offshore islands. After a few years, the rocks become covered with a coating of white guano which can be seen from some distance.

⇧

Their nests are separated exactly by the distance which an adult can reach, so that sitting birds and young are not exposed to the rapier-like bill thrusts of their neighbours when on the nest.

Cormorant

Phalacrocorax carbo

Broigheall

Cormorant colony in trees at sunset.

⇦

Adult cormorants in breeding plumage acquire a white patch on their flank and the gular pouch on the throat becomes white. This photograph shows a pair at the nest in an inland colony.

⇨

⇧

Cormorants nests in similar large colonies to gannets, but will also nest inland on tree-covered islands. They are a common sight on inshore waters, harbours and inland waterways. This photograph shows young cormorants near fledging, gathering at the edge of the colony waiting for their parents to bring food.

Shag

Phalacrocorax aristotelis

Seaga

⇧

Shags are like smaller versions of cormorants, however they show no white during the breeding season. Instead, they have a short erect crest. Unlike cormorants, shags nest in loose colonies, with individual nests being placed under boulders or in caves.

Grey heron

Ardea cinerea

Corr réisc

Grey herons nest almost exclusively in trees, building bulky nests at traditional sites. Nesting starts in February, accompanied by elaborate displays and much noisy calling. Here an adult heron raises breeding plumes as it settles onto its nest.

⇩

The grey heron is a common bird of inland waterways and coasts, generally to be seen standing silent, watching the water for any sign of movement. Their main food is fish, but they will take small mammals, insects, frogs and even young birds.

⇦

Because of their habit of occasionally taking young birds, herons are not always popular and are often driven away from a feeding area by intensive mobbing. Here the antagonist is a rook.

⇩

Mute swan

Cygnus olor

Eala bhalbh

Mute swans were introduced to Britain by the Romans and subsequently spread across to Ireland. Because of their elegance and beauty, they have been encouraged to nest on small waterways around towns and villages and have become a familiar sight throughout the country.

⇦

Bewick's swan

Cygnus columbianus

Eala Bhewick

Bewick's swans are scarce winter visitors to Ireland from their Arctic breeding grounds, occurring in large numbers at particular places, such as the Wexford Slobs.

⇨

Whooper swan

Cygnus cygnus

Eala ghlórach

Whooper swans are also winter visitors to loughs and reservoirs from their northern breeding grounds. They often feed in highly acid waters on peat bogs and their necks become stained a bright orange by the iron oxide in the water.

⇦

White-fronted goose

Anser albifrons

Gé bhánéadanach

White-fronted geese are another winter visitor to Ireland. Most Irish birds are of the race which breeds in Greenland. They winter on lowland peat bogs in Connemara, Donegal, Westmeath and the Wexford Slobs.

⇩

White-fronted goose, showing dark banding on the breast which indicates that it is an adult. The bill colour is typical of the race which breeds in Greenland.

⇦

Greenland white-fronted geese on pasture at Wexford.

⇩

Canada goose

Branta canadensis

Gé Cheanadach

Canada geese were introduced into this country to decorate ornamental ponds and lakes. They are a very successful species and have managed to colonise some of our natural waterways. Canada geese can be very aggressive during the breeding season and will attack anyone who approaches their nest.

Barnacle goose

Branta leucopsis

Gé ghiúrainn

Barnacle geese are winter visitors to offshore islands and some inland sites. They arrive in their thousands from their breeding grounds in Spitzbergen. Although they graze the grass very short and are reputed to do a lot of damage to pasture, they also enrich the soil with their droppings, providing a valuable source of fertiliser for island pastures.

Brent goose

Branta bernicla

Cadhan

Brent geese are much more restricted to coastal areas than most of the species of geese which occur in Ireland. Generally, they feed on sea grass (*Zostera*) beds on coastal mudflats, moving gradually up the shore as the tide rises, and also on coastal grasslands. Occasionally flocks of brent geese will graze football pitches because of the good quality grass which they support.

⇨

Shelduck

Tadorna tadorna

Seil-lacha

⇧

Strikingly marked, shelduck are a common sight on estuaries and bays. They nest in rabbit burrows and soon after hatching, their young gather to form large creches of several dozen young in the late summer near the breeding grounds.

Wigeon

Anas penelope

Lacha rua

The wigeon is one of our more striking ducks. The male has a bright chestnut head, yellow forehead and loud piping whistle, and the female is just as distinctive with her round dark chestnut head and pale rounded patch on her belly. In flight the male wigeon has a broad bright flash of white on his wings, making him easy to recognise from a distance.

⇦

Male (left) and female (right) wigeon feeding on the edge of mudflats, showing the difference between the plumages of the sexes.

⇨

Teal

Anas crecca

Praslacha

Teal are a common duck occurring on reed-fringed waterways and in muddy shallows. A few remain to breed in marshes and on lakes, while large numbers arrive in winter from their more northern breeding grounds.

⇨

The striking head pattern, white wing-flash and yellow triangle beneath the tail distinguish the male from the drab brown female teal.

⇩

Mallard

Anas platyrhynchos

Mallard

The male mallard is a striking sight with his bottle-green head and neck contrasting with the soft browns and greys of his body. The female is much less colourful, her black-streaked brown colours helping to camouflage her when on the nest to reduce the risk of being spotted by a predator.

⇦

Like all dabbling ducks, mallard feed by running their beaks through soft mud or up-ending to take vegetation from the bottom of pools, rivers and lakes.

⇨

A pair of mating mallard.

⇨

In the past, wild Mallard used to feed throughout the day, roosting at night. Because of hunting, they have changed their habits and generally loaf throughout the day in undisturbed sites, flying to nocturnal feeding grounds at dusk.

⇦

⇧

Mallard are our commonest resident duck and are regularly seen around rural towns and villages, where they interbreed with domestic duck to form strange hybrids. Here a brood of young, recently out of the nest, feed in a runnel in lakeside mud.

Shoveler

Anas clypeata

Spadalach

Shoveler are a common migrant to Ireland and breed in smaller numbers. They are distinguished from other ducks by their broad spatulate bill which they use to filter out food from the water and mud. The male can be recognised by his bottle-green head, chestnut patch over white on his flanks and black and white streaked back. Both sexes can be recognised in flight by their broad pale blue wing patch.

⇦

A male shoveler stands up on the water to stretch his wings, exposing his chestnut belly patch.

⇨

Pochard

Aythya ferina

Póiseard

Pochard are mainly a winter visitor from the north, although a few remain to breed. In winter, these birds form floating rafts on lakes and reservoirs, sometimes containing thousands of individuals.

Tufted duck

Aythya fuligula

Lacha bhadánach

Tufted duck are our most familiar diving duck. Unlike dabbling ducks, diving ducks submerge totally to feed on aquatic plants and insects from deeper waters. The small numbers of tufted duck which remain to breed are augmented in winter by huge flocks which migrate south from their northern breeding grounds to find more available food.

⇨

Hen harrier

Circus cyaneus

Cromán na gcearc

⇧

Like all harriers, hen harriers hold their wings in a distinct "V" while they quarter the ground searching for small rodents and birds such as meadow pipits. With its dove grey plumage and black wingtips, the male hen harrier is a striking sight.

The hen harrier is a scarce breeding bird of upland moors and bogs. In winter large numbers of hen harriers arrive in Ireland from northern breeding grounds. They sometimes form communal roosts. These are usually at old traditional sites, and often in company with other species of birds of prey, especially merlins.

⇦

Sparrowhawk

Accipiter nisus

Spioróg

⇧

The large forward facing eyes of diurnal birds of prey give them a good command of areas in front and to the side. This gives sparrowhawks an advantage as they hunt by flying low over the ground, rising only to clear obstacles, hoping to surprise feeding finch flocks.

This photograph is unusual as it shows a sparrowhawk with its intended prey - a starling - after they both struck a window and fell to the ground, dazed.

⇨

Although often seen, normally only a fleeting glimpse as the bird nips between hedgerows, sparrowhawks are a secretive bird. They build their nest high up in a fork of a tree deep in woodland.

⇩

Kestrel

Falco tinnunculus

Pocaire gaoithe

Kestrels nest in old, abandoned crows' nests in trees or on ledges on cliffs, quarries or derelict buildings. The young resemble the female, with barring over the wing coverts, a brown streaked head and a heavily barred brown tail.

⇩

⇧

In spite of their strong hooked bill and ferocious appearance, the parents can be surprisingly gentle as they tear up prey into strips which their small young can manage.

Kestrels are our most abundant bird of prey. They are a familiar sight hovering alongside roads and over farmland. The male can be distinguished from the female by its sparsely spotted chestnut back, grey head and grey tail with a broad black band.

⇨

As the young grow older, the parents begin to deliver whole animals as prey items. Here a female feeds a whole frog to its recently fledged youngster.

Merlin

Falco columbarius

Meirliún

Unlike most of our birds of prey, merlin populations appear still to be declining, partly due to loss of the wide undisturbed moors on which they nest. Here a juvenile merlin suns itself after being fed.

⇦

Merlins are our smallest birds of prey, but are very swift and powerful predators, occasionally taking birds as large as woodpigeons, which are bigger than the merlin itself. Here a male merlin brings a meadow pipit to its nest.

⇨

Peregrine

Falco peregrinus

Fabhcún gorm

Until recently, peregrine populations were declining and numbers reached an all-time low in the 1970s. In the last twenty years the population has recovered and most of the ancestral eyries are now being used again.

⇩

⇧

Peregrines are one of the fastest birds in the world, reaching speeds of up to 100 miles per hour as they dive or "stoop" on prey.

The peregrine is our largest breeding falcon. Like all falcons, the male is termed the tiercel and the female the falcon.

⇦

Pheasant

Phasianus colchicus

Piasún

⇧

Like various other game birds, pheasants were introduced into Ireland. However the natural habitat of the pheasant is in the bamboo and forest-covered hills of China. Here two males fight over the right to breed.

Hen pheasants are much less striking than their mates which helps them to avoid detection by predators when they are sitting on the nest or have small young.

⇨

Here a hen pheasant sits on a branch. Once mated, the male pheasant has no more to do with the female or young, leaving her to incubate and rear the young alone.

⇨

Water rail

Rallus aquaticus

Rálóg uisce

Water rails are fairly common, but because of their shy, secretive habits and good camouflage, they are rarely seen. Their pig-like grunting is more often heard coming from dense reedbeds. Their barred and streaked, brown and grey plumage helps to break up their outline against swamp vegetation and therefore makes them harder for predators to spot.

⇩

Corncrake

Crex crex

Traonach

Related to the water rail, the corncrake is similarly very secretive. However, when flushed or when stretching its wings like the bird in the photograph, the eye is drawn by the flash of bright orange on the bird's wings.

⇩

⇧

The corncrake used to be a common bird in rural areas. Rarely seen, their harsh creaking call, "crex crex", was a familiar part of summer for many people. Because of changes in agricultural practices, the corncrake has become a rare bird throughout Europe, probably one of Ireland's rarest breeding birds.

Moorhen

Gallinula chloropus

Cearc uisce

In the past, every village had a village pond and every village pond had at least one pair of moorhens. Now the birds have moved into towns to nest on ornamental lakes in parks and gardens.

Another member of the rail and crake family, the moorhen is one of the commonest birds in rural areas, with nesting pairs regularly spaced along ditches and rivers.

Oystercatcher

Haematopus ostralegus

Roilleach

⇧

An oystercatcher flies around the observer, calling to attract attention away from its nest or young.

A common bird of rocky coasts and estuaries, the oystercatcher draws attention to itself by its loud piping call and striking black and white plumage with orange bill. This photograph shows a flock in late winter. Some birds have already moulted into summer plumage, losing the black tip to the bill and the white neck band.

⇨

Ringed plover

Charadrius hiaticula

Feadóg chladaigh

⇧

An adult ringed plover moves onto its eggs to incubate. Like the adult, the eggs are patterned so that, when they are not covered, they cannot be easily seen against the stony background.

Like most waders, ringed plovers build virtually no nest, but lay their eggs in a shallow depression, scraping a few strands of vegetation and rabbit droppings together to make the eggs less obvious against the bare sand.

The young of most wading birds are able to walk within a few hours of hatching. Here a ringed plover chick, only a couple of hours out of the egg, walks over shingle away from the nest.

An immature ringed plover, almost ready to fly, has acquired the brown cap and black and white neck band which it will keep until it moults into adult colours to breed.

Golden plover

Pluvialis apricaria

Feadóg bhuí

A flock of golden plovers in the winter moving from the daytime loafing areas to feed at night in the fields.

⇩

Lapwing

Vanellus vanellus

Pilibín

⇧

Lapwings are a common sight throughout the year feeding in pasture and arable fields throughout the country, often seen flying up to noisily mob a predator or person who has approached too close to its nest.

In summer, the long curved crest of lapwings extends well beyond the back of their head and gives them a very easily recognised outline.

⇨

⇧

The broad rounded wings of the lapwing, with their distinctive black and white pattern, make it easily recognised in flight. Its familiarity in country areas has resulted in its having numerous alternative names such as pee-wit (which describes its call) and green plover.

Even in its downy juvenile plumage, the black collar and crown of the adult can already be seen.

⇩

Knot

Calidris canutus

Cnota

⇧

The knot is one of many species of wading bird which breeds in tundra and migrates south to winter on coastal mudflats. This photograph shows a knot in late summer, moulting from its breeding plumage of brick-red underparts and mottled silver and black back into its drab grey winter plumage.

A flock of knot on mudflats in winter. In sites with a rich inter-tidal invertebrate fauna, flocks of knot may number many thousand birds.

⇦

Sanderling

Calidris alba

Luathrán

Like the knot, sanderling do not breed in Ireland, but migrate south from Greenland and Spitzbergen to winter on our coasts. They can be recognised by their silvery white plumage and black bills as they race along the incoming tide. Here they are accompanied by four dunlins.

Little stint

Calidris minuta

Gobadán beag

Little stints occur regularly in small numbers on passage on Irish coastal mudflats and inland pools. A diminutive bird, its winter plumage is drab silvery grey. In summer it acquires the mottled wing pattern worn by the bird in the photograph.

Baird's sandpiper

Calidris bairdii

Gobadán Baird

Baird's sandpiper is one of many rare transatlantic vagrants which are brought to Ireland by strong winds and storms.

⇦

Dunlin

Calidris alpina

Breacóg

⇧

Dunlins are the most abundant wading bird on our shores, forming flocks of thousands which wheel over estuaries to roost as the tide rises. In summer dunlins have a black belly and rich golden brown upperparts worn by some of the birds in the photograph. Also in the picture are single sanderling and little stint.

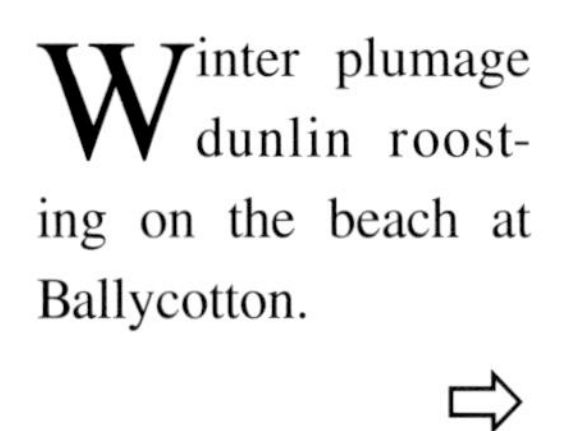

Winter plumage dunlin roosting on the beach at Ballycotton.

⇨

Snipe

Gallinago gallinago

Naoscach

⇧

Snipe are a common species occurring in marshes and bogs throughout the year. Their cryptic plumage affords them superb camouflage. They have such confidence in not being visible that they will remain still until almost trodden on. When they finally fly up, they will zig-zag away, calling loudly.

Woodcock

Scolopax rusticola

Creabhar

Unlike most of our wading birds, woodcock are a mainly nocturnal species, roosting through the day on the ground in deep woodland and flying at dusk to feed in fields and marshes.

⇩

Black-tailed godwit

Limosa limosa

Guilbneach earrdhubh

Black-tailed godwits in full summer plumage. In winter, they lose the rich red coloration which is replaced by light brown on the upperparts and a silvery belly.

Black-tailed godwits moulting into summer plumage at Lough Beg. Black-tailed godwits are common passage migrants and common winter visitors to our coasts and estuaries.

A flock of black-tailed godwits wheeling over the estuary to their roost as the tide rises.

Curlew

Numenius arquata

Crotach

⇧

The long, sensitive curved bill of the curlew is adapted to locate ragworms and lugworms deep in the mud of estuaries.

Curlew are the commonest of the large waders to occur on our coasts. In summer, they breed on moors and upland pasture, where their haunting, bubbling song is a frequent sound in spring and early summer. (The two smaller birds in the foreground are redshank.)

⇦

Greenshank

Tringa nebularia

Laidhrín glas

⇧

Greenshank are a regular passage migrant and winter visitor in small numbers to Irish coasts and inland waters, often feeding in gulleys and small pools on the upper part of mudflats.

Common sandpiper

Actitis hypoleucos

Gobadán

Common sandpipers breed on gravel and shingle spits on rivers and the edges of lakes throughout western Ireland. In the autumn they migrate to coastal areas and south as far as Africa to winter.

⇨

Unlike most waders, the common sandpiper can be found on almost any type of shore, from sand and mud to rocks and boulders or even concrete-lined ponds or reservoirs.

⇩

Turnstone

Arenaria interpres

Piardálaí trá

Winter plumage turnstones on a quay at high tide. Turnstones get their name from their habit of turning over stones, pebbles, clumps of seaweed and other tideline debris to catch invertebrates which hide underneath while the tide is out.

⇩

Black-headed gull

Larus ridibundus

Sléibhín

⇧

One black-headed gull bows to another in a dispute on the breeding grounds.

Of all our gull species, the black-headed gull is most likely to be seen in the company of rooks and jackdaws, following a tractor as it ploughs fields in autumn.

⇨

M D

⇧

In summer, in spite of their name, black-headed gulls acquire a dark brown hood, which is a major feature in its broad array of courtship displays.

Black-headed gulls are abundant throughout Ireland. In winter the dark brown hood is replaced by a mark behind the eye.

⇦

Common gull

Larus canus

Faoileán bán

⇧

Common gulls occur around our coasts in the winter, but breed in large colonies on lake islands in the northwest. They resemble herring gulls, which are our commonest gull, but are smaller, have a slim green or yellow bill, and a distinctive high mewing call.

Lesser black-backed gull

Larus fuscus

Droimneach beag

The lesser black-backed gull can be distinguished from the more common herring gull by its darker slate-grey back and yellow legs. It is more restricted to coastal areas than the herring gull, although both species will form large mixed colonies on cliffs and on coastal islands.

⇩

Herring gull

Larus argentatus

Faoileán scadán

⇧

An adult herring gull showing the silvery grey back and pink legs, standing among tussock of sea thrift *(Armeria maritima)* on a rocky sea cliff.

The herring gull is our commonest gull, breeding mainly on the coast. It is much more adapted to humans than other gulls and will even nest on rooftops in coastal towns. This species is equally at home on urban rubbish dumps as following trawlers far out to sea to scavenge scraps of the catch.

⇨

A young herring gull hatching. The red in the shell is all that is left of the yolk and the white spot on the tip of the bill is the egg tooth used by the chick to break through the tough eggshell and which drops off a few hours after hatching.

⇨

Young herring gulls are initially covered by a black-spotted grey down for camouflage, before they moult into their brown first-year plumage.

Great black-backed gull

Larus marinus

Droimneach mór

Great black-backed gulls are our largest breeding gull, occurring singly or in loose colonies on rocky headlands and islands. Part of their prey includes auks and shearwaters and the young of other birds.

⇦

A great black-backed gull arouses the concern of nesting cormorants as it scavenges through the colony.

⇨

Kittiwake

Rissa tridactyla

Saidhbhéar

⇧

Kittiwakes are the most truly oceanic of our breeding gulls. They nest on isolated cliffs and offshore islands, spending the winter months far out at sea.

Adult kittiwakes can be recognised by their small size, black legs and the "dipped in ink" wingtips.

⇨

Their nest is an untidy heap of seaweed lined with fine grass and rootlets which they build in large colonies on tiny ledges of steep cliffs.

⇩

Sandwich tern

Sterna sandvicensis

Geabhróg scothdhubh

⇧

Sandwich terns are our largest breeding terns, recognised by their all-black crest and yellow-tipped black bill. They nest in large dense colonies on shingle spits which they defend vigorously against all intruders.

Roseate tern

Sterna dougallii

Geabhróg rósach

Roseate terns are a scarce breeding bird on our coasts, occurring mainly in small numbers in large colonies of other tern species. They can be recognised by their mainly black bill, rose-flush to their breast and long tail streamers. The bird in the photograph has been ringed to monitor its movements.

⇦

Common tern

Sterna hirundo

Geabhróg

Common terns, like the other Irish species, nest on sand and gravel banks, laying their eggs in a hollow in the ground with no real nest. The young are marked with black spots on a brown background to camouflage them from predators.

⇨

Arctic tern

Sterna paradisaea

Geabhróg Artach

Arctic terns are very similar in appearance to common terns, but can be separated by their all blood-red bill and translucent underwings which can be clearly seen in this photograph. Arctic terns are generally much less widespread than common terns and are more frequently found nesting singly or in colonies of less than ten pairs on small headlands and offshore islands. They are very dependent upon the movements of sand-eels which are their main prey.

⇨

A pair of arctic terns calling near their nest.

⇦

Little tern

Sterna albifrons

Geabhróg bheag

Little terns are our smallest breeding tern, recognised in the breeding season by the white forehead and black-tipped yellow bill.

⇨

They nest in small colonies on beaches and as a result are often vulnerable to casual damage by bathers and other tourists who disturb the adults and trample nests.

⇨

Guillemot

Uria aalge

Foracha

Like diminutive penguins, guillemots are a striking member of our avifauna. They nest on steep coastal cliffs, where they lay their single egg on the bare rock, often on very narrow ledges. The egg is strongly pear-shaped, so that if its rolls, it does so in a circle and is less likely to fall off the nesting ledge.

⇨

In winter guillemots lose their chocolate brown mantle and are left with a black cap and back, and a dark line through the eye. When they leave the breeding ground, the young follow their parents out to sea and form large floating rafts offshore.

⇩

Razorbill

Alca torda

Crosán

⇧

Razorbills can be distinguished from guillemots in summer by their black and white plumage and broad white-striped bill. Unlike guillemots, razorbills nest in holes in cliffs and under rocks.

Puffin

Fratercula arctica

Puifín

Puffins are one of our best known seabirds. In summer, their harlequin-like bill and horn above the eye give them a comic, faintly quizzical look.

⇦

Unlike our other breeding auks, puffins nest in burrows deep in the soil on coastal headlands and rocky offshore islands. The edges of their bills are lined with razor-sharp serrations which help them to carry up to twenty sand-eels cross-wise, back to their nests.

⇩

Stock dove

Columba oenas

Colm gorm

⇧

Stock doves are a shy bird of rural areas and woodland. They nest in holes in old trees or in barns. They can be separated from woodpigeons by the lack of white on the neck and wings and two black bars on their wings.

Woodpigeon

Columba palumbus

Colm coille

Woodpigeons are common throughout Ireland, nesting in parks and gardens, woodland and hedgerows. On some offshore islands, where there are no trees, they will even nest on low banks or on the ground. Their gentle cooing is a familiar sound of summer.

⇨

A flock of woodpigeons taking off from a field of beans.

⇩

Collared dove

Streptopelia decaocto

Fearán baicdhubh

The collared dove was restricted to the Middle East until the first half of this century when it underwent a dramatic range expansion, spreading across the whole of Europe. It is not certain how they were able to colonise new areas, or why the range expansion was so fast, but it is likely that they are able to profit from increased agricultural intensification. Now collared doves are a familiar bird of towns and villages throughout the country.

⇨

Two juvenile collared doves almost ready to leave the nest.

⇦

Cuckoo

Cuculus canorus

Cuach

Cuckoos are parasitic upon other birds. The female lays her egg in the nest of a smaller bird such as a meadow pipit, and as soon as it hatches, the young cuckoo ejects the eggs of its host so that there is no competition for food. This photograph shows a recently fledged young cuckoo which will soon migrate south to Africa for the winter.

⇦

Barn owl

Tyto alba

Scréachóg reilige

Barn owls are probably the most abundant of Ireland's three breeding owl species. Often seen at dusk as they quarter hedges and ditches searching for voles, or sweeping up to their nest site in an old barn. Here an adult carries a dead rat back to its nest.

⇩

⇧

The barn owl returns with food to its hungry fledglings.

Long-eared owl

Asio otus

Ceann cait

During the day, long-eared owls roost against a tree trunk or in dense branches, raising their ear tufts to break up their outline and make them more difficult to see.

⇨

Long-eared owls nest in the old nest of other species of bird, particularly crows. They are strictly nocturnal and their presence is often only betrayed by their long quavering hoot.

⇩

Short-eared owl

Asio flammeus

Ulchabhán réisc

⇧

Short-eared owls fly mainly by day, although they will hunt at night when they have large young to feed. They hunt like a harrier, quartering the ground over marshes, bogs and pastures for voles, young rabbits and young birds.

Kingfisher

Alcedo atthis

Cruidín

⇧

The kingfisher is one of our most colourful birds. Rarely seen perched like this, it is more often seen as a flash of blue as it flies fast along a river.

Skylark

Alauda arvensis

Fuiseog

Skylarks are a common sight and sound in rural areas, where they hang in the air for hours pouring out their song over pasture and crops. They nest in a hollow in the ground and, within hours of hatching, the young leave the nest and separate to avoid detection by predators.

⇦

Sand martin

Riparia riparia

Gabhlán gainimh

Sand martins nest in long burrows in sand or gravel cliff faces, often in river banks and old sand pits. They can be recognised by their brown back and white underparts with a brown neck band.

⇨

Swallow

Hirundo rustica

Fáinleog

Swallows are one of the most familiar birds in rural areas. Most farms have at least one pair, nesting in a shallow mud cup attached to a beam or rafter. Only adult birds have the long tail streamers.

⇦

Meadow pipit

Anthus pratensis

Riabhóg mhóna

The meadow pipit is one of our most abundant birds. They nest on the ground in a shallow cup made of grasses and lined with hairs and roots. They are often seen in rough pasture, bogs, moors and coastal grassland.

⇨

Grey wagtail

Motacilla cinerea

Glasóg liath

⇧

In summer grey wagtails occur in similar habitats to dippers, liking fast rocky streams and mill races, building their nest on a cliff or low bank beside the river. In winter they can be seen in farmyards and in sub-urban areas.

Like pied wagtails, grey wagtails constantly wave their tails up and down, mimicking the movement of water and making them difficult to pick out in spite of their bright colours.

⇨

Pied wagtail

Motacilla alba

Glasóg shráide

Pied wagtails are less strongly associated with water than grey wagtails and are often seen in towns, feeding along streets and in town gardens. They nest in rock crevices in walls and buildings. The bird in the photograph is carrying a bundle of insects back to the nest for her young.

⇩

Pied wagtails roosting in an urban tree.

⇨

Waxwing

Bombycilla garrulus

Síodeiteach

The waxwing is a scarce winter visitor to Ireland. In most years there will be only one or two records. However occasionally, due to certain conditions on the Scandinavian breeding grounds, there will be an "eruption" and large numbers will arrive to feed on berries, often in parks and gardens.

⇨

Dipper

Cinclus cinclus

Gabha dubh

⇧

Dippers are shy birds of rocky rivers and fast flowing streams. They are like a large wren, but with chocolate brown plumage contrasting with a clean white breast. As a nest they build a large ball of moss and grass in an overhanging cliff or under a bridge. Another favourite nest site is on a cliff behind a waterfall.

Wren

Troglodytes troglodytes

Dreoilín

Wrens are one of the most abundant and widespread birds in Ireland, occurring from urban gardens, to the wildest sea cliffs and high mountains. In spite of its small size, the wren has a loud ringing song which can be heard throughout the year, even in the wildest weather.

⇦

The male wren builds a number of nests in ivy hanging from old fence posts, small cliffs and banks or the hole of an old tree. Once the female has chosen the one which she thinks is suitable, the male will line it with feathers and it will be used to rear their young. The other nests are then used as roost sites and are termed cock's nests.

⇩

Robin

Erithacus rubecula

Spideog

Like the wren, robins are another familiar sight in gardens and parks, often becoming so tame that they may even perch on the blade of a fork as it is used to turn the soil. In winter, each individual robin is highly territorial; females and males will even drive out their mate from the previous season. When spring comes however, they overcome this territoriality and pairs join forces to drive out any potential intruders.

⇨

In the autumn, robins change their song and produce a much thinner, sadder tune until the weather starts to warm up again. Unless they are helped by people putting out food, or turning over soil, winter can be a hard time for robins as they must take fruit and berries to replace their summer food of insects.

⇩

Stonechat

Saxicola torquata

Caislín cloch

The close relationship between robins and stonechats can be recognised by their similarity in shape and their habit of flicking their tail when alarmed. The male stonechat can be distinguished from the female by its black head. Both sexes share the bright orange breast.

⇦

Stonechats are often seen standing on an elevated perch such as a fence post, or the top of a bush, scanning the ground for the movement of beetles and other insects or chakking loudly as the observer enters their territory.

⇨

Wheatear

Oenanthe oenanthe

Clochrán

⇧

The wheatear is another chat, closely related to both stonechat and robin. Unlike these species, it is a summer migrant to our shores, wintering in Africa and the first birds arriving to set up territory on the breeding grounds in late February. Wheatears nest in a hole between stones in old stone walls, in old rabbit burrows or in cliff faces. The male can be distinguished by its grey back and black ear-patches.

Blackbird

Turdus merula

Lon dubh

⇧

A male blackbird at the nest.

Occasionally, a blackbird will appear which is mainly white. This trait, termed "albinism", is genetic and the genes for white feathers are passed down from parent to offspring.

⇨

Blackbirds must be our most familiar bird. Most gardens have a resident pair and many will have a nest each year in a hedge or low bush. Throughout the summer, blackbirds feed themselves and their young on insects and worms. In the autumn, they profit from the abundance of fruit to gorge themselves on berries to build up a store of fat for the winter.

⇦

Fieldfare

Turdus pilaris

Sacán

Like redwings, fieldfares breed mainly in Scandinavia and migrate south in the autumn to avoid the heavy snows and very low temperatures on the breeding grounds. They are our most brightly coloured thrush with a striking grey head and rump, black-tipped yellow bill and dark spotted peach-coloured breast.

⇦

Song thrush

Turdus philomelos

Smólach ceoil

Song thrushes are another familiar sight in parks and gardens. Unlike blackbirds which build a bulky nest entirely composed of grasses, roots and a few hairs, song thrushes line their nest with mud. They lay four or five sky-blue eggs with small black spots. Song thrushes can be very beneficial to gardeners, as they take large numbers of snails and slugs.

⇨

Redwing

Turdus iliacus

Deargán sneachta

Redwings are a member of the thrush family which breeds in Scandinavia and migrates to Ireland in the winter, arriving in autumn to compete with our resident song thrushes, blackbirds and mistle thrushes for berries. Redwings can be identified by their bright white eyestrip and flash of red under their wings.

⇨

Mistle thrush

Turdus viscivorus

Liathráisc

Mistle thrushes are a typical bird of open landscapes with large trees. They nest early in the year, building a large nest of sticks and mud, normally in the fork of a large tree. Occasionally, like the bird in the photograph, they will nest in buildings.

⇦

Sedge warbler

Acrocephalus schoenobaenus

Ceolaire cíbe

⇧

Sedge warblers build their nests low down near the base of rank herb vegetation close to marshes. Their close relatives, reed warblers, build their nests hanging between the stems of reeds, each nest only supported by a few strands of woven grasses.

Sedge warblers are common in marshes and reed-fringed ditches. They return from their African wintering grounds in late April or early May and announce their presence as they establish their territories by a loud scratching song delivered from bramble bushes or in a rapid fluttering song flight. They can be distinguished from the closely related reed warbler by their broad white eyestrip and pattern of black streaking on the back.

⇨

Reed warbler

Acrocephalus scirpaceus

Ceolaire giolcaí

⇧

The reed warbler has only recently become established as a breeding species in Ireland. Like sedge warblers, reed warblers migrate south for the winter. They are very dependent upon aphids which feed on common reed (*Phragmites australis*) which they scoop up in huge quantities to build up fat for the long migration. Reed warblers are much more restricted to true reed marsh than sedge warblers, although they collect much of the food for their young from adjacent stands of willows.

Chiffchaff

Phylloscopus collybita

Tiuf-teaf

The chiffchaff is like a browner version of the willow warbler and can only be separated conclusively by its darker legs and different song. Chiffchaffs are mainly summer migrants. However a small number overwinter.

⇨

Willow warbler

Phylloscopus trochilus

Ceolaire sailí

Willow warblers and their close relatives chiffchaffs are some of the first migrants to return to our shores in the spring. The monotonous song of the chiffchaff can be heard long before the leaves are fully open. Willow warblers arrive a little later and can be heard singing a melodious descending whistle. Both species build a ball of dried grasses in which to nest. Willow warbler nests tend to touch the ground, while chiffchaffs build in vegetation just above ground level.

⇦

Spotted flycatcher

Muscicapa striata

Cuilire liath

⇧

Spotted flycatchers occur mainly in areas of parkland or on the edge of mature woodland, where they nest in a hollow in the bark of an old tree or in ivy. They are most often seen sitting on a low post or branch and making sorties into the open to catch insects with an audible snap of the beak.

Spotted flycatcher at the nest.

⇦

Long-tailed tit

Aegithalos caudatus

Meantán earrfhada

⇧

Long-tailed tits are very sociable birds, roosting in lines on branches, huddled together to keep warm outside the breeding season. They are almost invariably seen in flocks, often accompanied by other species of bird. They even help each other to build nests. Each nest is composed entirely of lichens, mosses and feathers, held together with spiders' web. Even with help, it may take a pair a month to complete their nest.

Coal tit

Parus ater

Meantán dubh

The coal tit is our smallest tit. It can be identified by the white line on its nape and two dotted white wingbars. Coal tits are normally associated with coniferous trees, particularly firs. This bird is carrying a white moth back to feed its chicks.

⇨

Blue tit

Parus caeruleus

Meantán gorm

Blue tits are a familiar sight in towns and gardens, hanging acrobatically from the bird table to extract peanuts, or taking the tops off milk bottles on the doorstop early in the morning. In the summer, they can be an enormous help to gardeners and farmers as they feed themselves and their young almost exclusively on caterpillars.

⇩

Great tit

Parus major

Meantán mór

Similar to blue tits, great tits are much bigger, with a black cap and a black stripe running right down the centre of his belly to the tail. Great tits tends to take different food to blue tits and often when you hear a tapping sound in the woods, it's not a woodpecker, but a great tit trying to open a nut which it has wedged in a crevice in the bark, or is holding with one foot.

⇦

Treecreeper

Certhia familiaris

Snag

Treecreepers are a common bird of woodland, parks and small copses. However because of their shy habits and good camouflage, their presence is often only betrayed by their high-pitched call. They roost in crevices in the bark of trees and are normally seen climbing up or down the bark of trees searching for insects.

⇨

Magpie

Pica pica

Snag breac

⇧

The magpie is an abundant and familiar bird. Its black and white plumage, long black tail and its habit of standing on the top of a bush or tree attracts attention. Magpies build large domed nests of thick twigs deep in the centre of thorn trees, so that their young and eggs are not taken by predators.

Chough

Pyrrhocorax pyrrhocorax

Cág cosdearg

Choughs are our rarest member of the crow family. They are very similar to jackdaws in their flight and actions, but their striking curved red bill and loud calls separate them from all other species of crow.

⇦

They are found almost exclusively on wild rocky sea cliffs, although a few pairs nest in old mine entrances and quarries in the mountains. The pair in the photograph are feeding young which are nearly ready to leave the nest and resemble the adults except that they have not fully acquired the long curved bill which helps them to extract insects from sheep-grazed turf on cliff-tops.

⇨

Jackdaw

Corvus monedula

Cág

Jackdaws are a familiar sight around human habitation, nesting in chimney pots and under the eaves. Their grey neck patch and striking pale blue eyes distinguish them readily from our other crow species. Here a small flock land on the back of a cow to pluck its hair in order to line their nests.

⇨

An adult jackdaw entering its nest in an old broken drainpipe. Most of the food which it brings back to its young is carried in its crop until the young are almost ready to leave the nest.

⇦

Jackdaws build large bulky nests of sticks lined with soft fur or feathers, which they sometimes collect before the previous owner has entirely finished with them.

⇨

Rook

Corvus frugilegus

Rúcach

Rooks are very sociable birds, rarely seen alone. They nest in colonies in tall trees and feed in flocks on arable land. They can be identified by the long baggy feathers on their legs, rounded tail and by the bare base to their bill which gives them a white-face appearance.

⇨

Hooded crow

Corvus corone corvix

Feannóg

The hooded crow is a common bird, found throughout rural areas, building a large bulky stick nest in trees. They are generally scavengers feeding along the tideline or, like the bird in the photograph, taking road-killed animals.

⇦

Raven

Corvus corax

Fiach dubh

⇧

The raven is the largest member of the crow family. They are generally shy of man and are mainly found away from human habitation, nesting in tall trees or on coastal cliffs. They build their nests in February and sometimes a late fall of snow will cover the sitting female. In spring, the male performs a spectacular display, rolling and tumbling in flight as he calls.

Raven calling against the sunset.

⇨

Starling

Sturnus vulgaris

Druid

The starling is one of our most familiar birds, nesting under the eaves of urban and rural houses. The song of the starling is very varied, being composed of scratching whistles, interspersed with mimicry of other birds and familiar sounds.

⇦

⇧

A starling landing at the nest.

In summer, starlings lose much of the spotting over their head and neck and, for a while at the height of the breeding season, the bill of the male takes on a blueish hue. The starling in this photograph is collecting nesting material.

⇨

In the winter, starlings form huge roosts and the sight of a flock wheeling through the air with military precision to roost is breathtaking.

⇦

House sparrow

Passer domesticus

Gealbhan binne

The house sparrow is another very familiar bird of towns and villages. Building bulky nests of straw or grasses under the eaves or occasionally in trees, their chirping is a characteristic sound of lazy summer days.

⇨

Tree sparrow

Passer montanus

Gealbhan crainn

Unlike its close relative the house sparrow, tree sparrows are much less attached to human habitation, nesting in holes in trees or walls around farms. They are rarely seen in towns or cities. In Ireland, they are found in small colonies, mainly around the coast, where they nest in crevices of derelict buildings. They can be separated from house sparrows by the all-brown cap and a dark spot like a question-mark on their cheek.

⇦

Chaffinch

Fringilla coelebs

Rí rua

⇧

A bird of woodland, parks and gardens, the chaffinch can often be found in the spring singing its loud descending song from a raised perch. Chaffinches build a delicate nest of moss and lichen lined with horse hair in the fork of a tree. The female is much more drab than the male shown in the photograph, being overall brown and black in colour, with flashes of white on its outer tail feather and wings.

Goldfinch

Carduelis carduelis

Lasair choille

The goldfinch is one of our more striking small birds, with its red and white face-pattern, yellow flashes on the wings and melodic twittering call. Goldfinches are often seen feeding on dead thistle heads, from which they take seeds. In the autumn and winter, goldfinches form large flocks with other species of finch to feed on seeds in arable fields.

⇨

Mixed finch flock with goldfinches, linnets and bramblings (*Fringilla montifrigilla*), a winter visitor from Scandinavia.

⇨

Siskin

Carduelis spinus

Siscín

Siskins nest in conifers, often in spruce plantations, where they build a shallow cup, high in the branches. In winter, they often associate with redpolls and tits, often feeding in alders.

⇨

The female siskin in the photograph can be separated from the male by its drab brown upperparts, without the bright black cap and yellow breast of the male.

⇦

Linnet

Carduelis cannabina

Gleoiseach

Linnets are abundant in summer wherever there is gorse, building their cup nest of grasses in a low fork, protected from danger by the long spikes of the gorse.

⇩

Redpoll

Carduelis flammea

Deargéadan

The redpoll is an uncommon breeding bird, usually nesting in alders along river banks. Their numbers are augmented in the winter by migrants from the north and these can be seen throughout the winter in large flocks, often associated with siskins.

⇦

This photograph shows a female redpoll at the nest.

⇨

Bullfinch

Pyrrhula pyrrhula

Corcrán coille

Bullfinches are a common bird of gardens and hedgerows. They build a surprisingly thin nest of grass and rootlets in a bush or small tree.

⇦

The male bullfinch can be separated from the female by his bright pinky-red breast. In spite of their size, bullfinches are a shy, rarely seen bird. Their gentle piping call is often the only indicator of their presence.

⇨

Yellowhammer

Emberiza citrinella

Buíóg

Yellowhammers are a familiar bird of rural areas, singing their "little bit of bread and no cheese" from the tops of laneside bushes or coastal gorse. The bright yellow of the male's head is replaced in the female by brown. Yellowhammers build a bulky nest of grasses near the ground at the base of a hedge or bush, where the female lays brownish eggs patterned with scrawls and spots of black.

⇨

Reed bunting

Emberiza schoeniclus

Gealóg ghiolcaí

Throughout most of the year reed bunting are found in damp marshy areas, the males often drawing attention to themselves by sitting on the tops of marsh plants and flicking their tail with its bright white outer feathers. The male reed bunting loses his spectacular black cap and chestnut back in the winter and resembles the drab brown female.

⇦

English Index

Latin Index

Irish Index